Bibliographic information published by the German National Library:

The German National Library lists this publication in the National Bibliography; detailed bibliographic data are available on the Internet at http://dnb.dnb.de .

Imprint:

Print and binding: Books on Demand GmbH, Norderstedt Germany
ISBN: 9783668543744

This book at GRIN:

http://www.grin.com/en/e-book/376865/solar-pv-panels-efficiency-enhancement-fixed-and-tracking-system-and-energy

Chirag Anilkumar

Solar PV Panels Efficiency Enhancement. Fixed and Tracking System and Energy Value

A Comparative Study

GRIN Publishing

Solar PV Panels Efficiency Enhancement. Fixed and Tracking System and Energy Value

A comparative study

Abstract

The study conducts a comparison based on net efficiency and economic costs involved between tracking and fixed solar systems, so as to analyze and interpret the data for developing sustainable solutions for tapping solar energy through calculation of energy value produced per year under optimum conditions analyzing the data obtained through PVWatts calculator available at NREL

Table of contents

1. Introduction

Global warming has increased the demand and request for green energy produced by renewable sources such as solar power. Solar panels are used to convert light energy into electrical energy. Jawaharlal Nehru National Solar Mission (JNNSM) has set the ambitious target of deploying 1 lakh MW of grid connected solar power by 2022.

Doubling the share of solar, wind and other renewable sources in the energy mix would boost global economic growth by as much as 1.1 % by 2030, according to IRENA (International Renewable Energy Agency). The gains equal to $1.3 trillion would mostly come from increased investment.

The earth receives 16 x 10^{18} units of energy from the sun annually, which is 20,000 times the requirement of mankind on the Earth. In today's time solar cells are becoming extremely popular for utilizing solar energy to use in different ways such as producing electricity, transportation, cooking etc .So many solar panels have been installed around the world, acting as sustained sources of renewable energy

1.1. Problem Analysis

Solar panels are installed in the direction of maximum radiation of sunlight but the problem arises due to the fact that the sun is moving .So we cannot harness maximum radiation of sun all the time as the position of maximum radiation comes only once in 24 hours. Solar tracking is increasingly being applied as a sustainable power generating solution. The optics in concentrated solar applications accepts the direct component of sunlight light and therefore must be oriented appropriately to collect energy. It is done by moving the solar panel according to the movement of sun we can always receive the maximum radiation.

Solar Tracking System

Solar tracking system is a device for orienting a solar panel or concentrating a solar reflector or lens towards the sun. Concentrators, especially in solar cell applications, require a high degree of accuracy to ensure that the concentrated sunlight is directed precisely to the powered device.

Direct radiation is used to describe solar radiation traveling on a straight line from the sun down to the surface of the earth. Diffused radiation describes the sunlight that has been scattered by molecules and particles in the atmosphere but that has still made it down to the surface of the earth.

The energy contributed by the direct beam drops off with the cosine of the angle between the incoming light and the panel. In addition, the reflectance (averaged across all polarizations) is approximately constant for angles of incidence up to around 50°, beyond which reflectance degrades rapidly

The use of lens or mirror increases the tracker's efficiency since large amount of sunlight is concentrated on the panel and large power is generated. It can also reduce the size of the solar cell required to generate large power. It also has high optical efficiency.

.Precise tracking of the sun is achieved through systems with single or dual axis tracking.

Single Axis Tracking

For solar panels with single axis tracking systems, the panels can turn around the centre axis. The single axis tracking system consists of two LDR's placed on either side of the panel. Depending on the intensity of the sun rays one of the two LDR's will be shadowed and the other will be illuminated. The LDR with the maximum intensity of the sun's radiation sends stronger signal to the controller which in turn sends signal to the motor to rotate the panel in the direction in which the Sun's intensity is of maximum value.

Single axis tracker can only track the daily movement of the sun and not the yearly movement. The efficiency of the single axis tracking system is also reduced during cloudy days since it can only track the east-west movement of the sun.

Dual Axis Tracking

Dual axis tracking is typically used to orient a mirror and redirect sunlight along a fixed axis towards a stationary receiver. But the system can also gain additional yield on your PV cells.

Dual axis tracking system uses the solar panel to track the sun from east to west and north to south using two pivot points to rotate. The dual axis tracking system uses four LDR's, two motors and a controller. The four LDR's are placed at four different directions. One set of sensors and one motor is used to tilt the tracker in sun's east - west direction and the other set of sensors and the other motor which is fixed at the bottom of the tracker is used to tilt the tracker in the sun's north-south direction. The controller detects the signal from the LDR's and commands the motor to rotate the panel in respective direction.

Proposed Method

It is well known that if you compare identical arrays, one fixed and the other tracked, that the tracked array will annually outperform the fixed array.

Using related data, costs involved and theoretical values, tracking solar panels can be compared to static solar panels. By proper analysis using theoretical and calculated data available the efficiency of tracking systems can be brought out in terms of reduced payback period through calculation of energy value produced per year and efficient power generation bringing out the cost effectiveness of tracking systems and its related contribution to sustainable energy production.

2. Efficiency Analysis

Solar energy received per day for every month is calculated and its subsequent energy value is analysed using the online software available at NREL (PV Watts calculator- http://pvwatts.nrel.gov).

For Photovoltaic System, power can be calculated via

$E = A \times r \times H \times PR$

Energy Value is the system's total electrical output for the year multiplied by the retail electricity price, assuming an annual degradation rate of 0.5% from the Year 1 value:
Energy Value = Energy (kWh) × Average Cost of Electricity Purchased from Utility (₹/kWh)

Resultant output is as follows considering an example coastline city for comparison:

Location and Station Identification	
Requested Location	8.5074° N, 76.9730° E
Weather Data Source	(INTL) TRIVANDRUM, INDIA Range (4.0 km)
Latitude	8.48° N
Longitude	76.95° E

PV System Specifications	
DC System Size	4 kW
Module Type	Thin Film
Array Type	Fixed (open rack)
Array Tilt	10°
Array Azimuth	180°
System Losses	14%
Inverter Efficiency	96%
DC to AC Size Ratio	1.1
Initial Economic Comparison	
Average Cost of Electricity Purchased from Utility	3.63 ₹/kWh
Initial Cost	127.38 ₹/Wdc
Cost of Electricity Generated by System	5.90 ₹/kWh

Month	Solar Radiation (kWh / m^2 / day)	AC Energy (kWh)	Energy Value (₹)
January	6.20	595	2,160
February	6.70	579	2,102
March	6.83	657	2,384
April	6.04	564	2,049
May	5.47	531	1,929
June	4.63	438	1,592
July	4.94	483	1,753
August	5.44	531	1,926
September	5.84	552	2,003
October	5.63	546	1,983
November	5.32	498	1,809
December	5.69	548	1,990
Annual	5.73	6,522	₹ 23,680

PV System Specifications	
DC System Size	4 kW
Module Type	Thin Film
Array Type	1-Axis Tracking
Array Tilt	8.5°
Array Azimuth	180°
System Losses	14%
Inverter Efficiency	96%
DC to AC Size Ratio	1.1
Ground Coverage Ratio	0.4
Initial Economic Comparison	
Average Cost of Electricity Purchased from Utility	3.63 ₹/kWh
Initial Cost	127.38 ₹/Wdc
Cost of Electricity Generated by System	5.07 ₹/kWh

Month	Solar Radiation (kWh / m^2 / day)	AC Energy (kWh)	Energy Value (₹)
January	7.21	695	2,524
February	8.00	694	2,520
March	8.28	800	2,902
April	7.07	663	2,407
May	6.36	621	2,254
June	5.16	491	1,781
July	5.59	549	1,993
August	6.23	610	2,214
September	6.76	640	2,324
October	6.42	624	2,265
November	6.05	569	2,065
December	6.59	638	2,317
Annual	6.64	7,594	₹ 27,566

PV System Specifications	
DC System Size	4 kW
Module Type	Thin Film
Array Type	2-Axis Tracking
Array Tilt	8.5°
Array Azimuth	180°
System Losses	14%
Inverter Efficiency	96%
DC to AC Size Ratio	1.1
Initial Economic Comparison	
Average Cost of Electricity Purchased from Utility	3.63 ₹/kWh
Initial Cost	127.38 ₹/Wdc
Cost of Electricity Generated by System	4.82 ₹/kWh

Month	Solar Radiation (kWh / m^2 / day)	AC Energy (kWh)	Energy Value (₹)
January	7.80	750	2,722
February	8.39	728	2,641
March	8.52	824	2,989
April	7.30	684	2,485
May	6.80	663	2,407
June	5.57	529	1,920
July	6.01	589	2,138
August	6.50	636	2,310
September	6.90	654	2,373
October	6.57	638	2,316
November	6.36	596	2,164
December	7.20	695	2,524
Annual	6.99	7,986	₹ 28,989

Applying the above procedure, necessary data can be evaluated for various coastline places of India, such as Trivandrum, Chennai, and Mangalore.

Graph can be plotted against the energy value obtained per year according to the optimum power rating and cost of electricity generated according to the state tariff, to the variation in latitude in these places, for each of fixed, single-axis tracking and double-axis tracking.

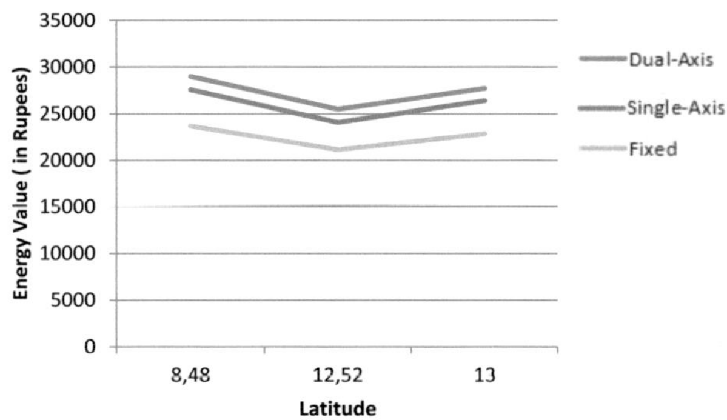

3. Payback Period

Following consideration can be applied during the calculation of the payback period

- Cost of the PV solar panels (with included battery storage) = Rs. 2 lakh
- Electronic piston cylinder (with support) to move the panel = Rs 10,000
- Electronic circuits with sensors to orient and control movement = Rs.2000

Latitude (^{0}N)	Fixed (years)	Single-Axis(years)	Double-Axis(years)
8.48	9	8	8
12.52	10	9	9
13	9	8	8

Inferences

From the above data and calculations obtained, following inferences can be made:

- There is substantial increase in the energy value when fixed type solar panels are replaced by single-axis or double-axis tracking systems
- As energy production and its projected payback period is similar for dual and single-axis tracking, it's economical to use dual axis trackers are typically used in smaller residential installations and locations with very high government feed in tariffs
- Tracking systems are more affective in large scale commercial systems more than residential systems
- Change in energy production depends mainly on change in latitude, geographical and climatic factors
- Efficiency of tracking systems is identified in terms of the energy value obtained per year and analyzed through reduced payback period
- There is an almost steady change in the energy value produced per year for various setups of PV panels
- Near the equator, one will have the highest benefit of tracking the sun.
- The space requirement for a solar park is reduced, and they maintain the same output
- The use of lens or mirror increases the tracker's efficiency since large amount of sunlight is concentrated on the panel and large power is generated

Acknowledgment

I thank Aneez Hydari, Associate Professor, Government Engineering College Thrissur for supporting and guiding me throughout this study.

References

Ashraf Balabel, Ahmad A et al- "Design and Performance of Solar Tracking Photovoltaic System; Research and Education", INTERNATIONAL JOURNAL OF CONTROL, AUTOMATION AND SYSTEMS, VOL1 NO.2

Chirag Anilkumar –"Feasibility Study of Hybrid Renewable Energy System in Kerala", ID-352, Energy Systems, KETCON 2016

Dhanabal.R, Bharathi et al -"Comparison of Efficiencies of Solar Tracker systems with static panel Single-Axis Tracking System and Dual-Axis Tracking System with Fixed Mount ",International Journal of Engineering and Technology (IJET)

National Renewable Energy Laboratory *(*www.nrel.gov)

Ministry of New and Renewable Energy *(*www.mnre.gov.in)

Vanjari Venkata Ramana and Debashisha Jena, "An accurate modeling of photovoltaic system for uniform and non-uniform irradiance", International Journal of Renewable Energy Research, Vol.5, No.1, 2015)